Written by: Nigel Norvell
Illustrated by: Olena Oprich

This is a work of fiction. Names, characters, places, and events are the product of the author's imagination or are used in a fictitious manner. Any resemblance to actual persons, living or dead, or actual events is purely coincidental.

ISBN: 979-8-9940100-1-3
First Edition
Printed in the United States of America

# MILITARY KIDS SERIES:
## ABCs

Author
Nigel Norvell

Illustrator
Olena Oprich

# A AIRBORNE

Airborne soldiers lead the way.

# B BOOTS

Boots get tied tight every day.

# C COMPASS

Compass points north, south, east, and west.

# D DOG TAGS

Dog tags are worn around the neck.

# E ENGINEER

Engineers build bridges to help us go.

# F FIGHTER JET

Fighter jets zoom fast—enjoy the show!

# G GEAR

Gear keeps us ready, packed up tight.

# H HELICOPTER

Helicopters hover up in the sky.

# I INSPECTION

Inspections require rulers to measure.

# J JUMP

Jump—keep feet and knees together.

# K K9

K9s train loyal, strong, and true.

# L LEADERS

Leaders guide the team in all we do.

# M MREs

MREs can be a wonderful delight.

Meal, Ready-to-Eat

MRE

MENU 10
Chili and
Macaroni

CRACKERS

# N

NVGs help you see at night.

# O OBSTACLE COURSE

Obstacle courses will help you train.

# P PILOT

Pilots fly high in their plane.

# Q QUARTERMASTER

Quartermasters bring the gear with speed.

# R RUCKSACK

Rucksacks carry the gear we need.

# S SUBMARINE

Submarines dive silent through the ocean.

# T TANK

Tanks rumble forward in steady motion.

# U UNIFORM

Uniforms show pride and discipline too.

# V  VICTORY

Victory comes when the mission is through.

# W

## WATER

Water fills the green canteen

# X X-RAY

X-rays check bones for you and me.

# Y YAWN

Yawn in the morning to start the day.

# Z ZERO

"Zero, zero, zero!"—the grader won't go away.

www.ingramcontent.com/pod-product-compliance
Lightning Source LLC
Chambersburg PA
CBHW041556040426

42447CB00002B/187